T0281113

Cambridge Elements ≡

Elements in Geochemical Tracers in Earth System Science
edited by
Timothy Lyons
University of California
Alexandra Turchyn
University of Cambridge
Chris Reinhard
Georgia Institute of Technology

BARIUM ISOTOPES

Drivers, Dependencies, and Distributions through Space and Time

Tristan J. Horner
Woods Hole Oceanographic Institution
Peter W. Crockford
Weizmann Institute of Science

CAMBRIDGE
UNIVERSITY PRESS

CAMBRIDGE
UNIVERSITY PRESS

University Printing House, Cambridge CB2 8BS, United Kingdom

One Liberty Plaza, 20th Floor, New York, NY 10006, USA

477 Williamstown Road, Port Melbourne, VIC 3207, Australia

314–321, 3rd Floor, Plot 3, Splendor Forum, Jasola District Centre,
New Delhi – 110025, India

79 Anson Road, #06–04/06, Singapore 079906

Cambridge University Press is part of the University of Cambridge.

It furthers the University's mission by disseminating knowledge in the pursuit of
education, learning, and research at the highest international levels of excellence.

www.cambridge.org
Information on this title: www.cambridge.org/9781108791113
DOI: 10.1017/9781108865845

First published 2021

A catalogue record for this publication is available from the British Library.

ISBN 978-1-108-79111-3 Paperback
ISSN 2515-7027 (online)
ISSN 2515-6454 (print)

Barium Isotopes

Drivers, Dependencies, and Distributions through Space and Time

Elements in Geochemical Tracers in Earth System Science

DOI: 10.1017/9781108865845
First published online: March 2021

Tristan J. Horner
Woods Hole Oceanographic Institution

Peter W. Crockford
Weizmann Institute of Science

Author for correspondence: Tristan J. Horner, tristan.horner@whoi.edu

Abstract: In the modern marine environment, barium isotope ($\delta^{138}Ba$) variations are primarily driven by barite cycling – barite incorporates "light" Ba isotopes from solution, rendering the residual Ba reservoir enriched in "heavy" Ba isotopes by a complementary amount. Since the processes of barite precipitation and dissolution are vertically segregated and spatially heterogeneous, barite cycling drives systematic variations in the barium isotope composition of seawater and sediments. This Element examines these variations; evaluates their global, regional, local, and geological controls; and, explores how $\delta^{138}Ba$ can be exploited to constrain the origin of enigmatic sedimentary sulfates and to study marine biogeochemistry over Earth's history.

Keywords: barite, biogeochemistry, paleoceanography, marine sulfate, productivity proxy

ISBNs: 9781108791113 (PB), 9781108865845 (OC)
ISSNs: 2515-7027 (online), 2515–6454 (print)

Contents

1 Introduction

Barium (Ba) and its isotopes are, at face value, an unusual choice for a biogeochemical proxy: Ba has no known biochemical role, is not appreciably accumulated by organic matter, and, being a heavy element – its name derived from the Greek *barys* meaning heavy – is not expected to exhibit large isotopic variations in the environment. However, as we show in this Element, the cycling of Ba and its isotopes are intimately connected with marine biogeochemistry. These connections form the basis of two widely used paleoceanographic proxies for marine nutrients and carbon export. That these same connections render significant isotopic effects means that Ba isotopes can also be leveraged to provide novel genetic constraints on the origin of enigmatic sedimentary sulfates deposited in the ancient geological record, and to interrogate key aspects of marine biogeochemistry through time.

In the modern marine environment, Ba exhibits two curious correlations that underpin its use as a paleoceanographic proxy. First, the concentration of dissolved Ba in seawater, hereafter [Ba], is correlated with the concentration of the major nutrient silicon (as silicate, hereafter [Si]; Chan et al., 1977; Figure 1). Second, the rate of accumulation of particulate Ba in certain marine sediments is positively correlated with the downward flux of organic carbon (e.g., Eagle et al., 2003; Paytan & Griffith, 2007; Yao et al., 2020). Assuming that both of these relationships held true across much of Earth's past, the abundance of Ba in certain types of sediment can be used as a proxy of the nutrient content of past seawater or of the strength of the biological carbon pump.

Unlike the major nutrients, however, the principle dissolved–particulate transformation of Ba is not driven by uptake into marine phytoplankton, but rather by precipitation into discrete, micron-sized grains of the mineral barite ($BaSO_4$, barium sulfate; Dehairs et al., 1980). This is paradoxical, in that the ocean is mostly undersaturated with respect to $BaSO_4$ (see Box 1), implying that additional processes are necessary to drive $BaSO_4$ precipitation from seawater. The most widely accepted solution to the $BaSO_4$ paradox is the microenvironment-mediated model of precipitation, first proposed by Chow and Goldberg, 1960: During microbial oxidation of sinking organic matter, larger aggregates may develop chemically distinct microenvironments in which respired Ba^{2+} and SO_4^{2-} ions accumulate. Eventually, some microenvironments become sufficiently supersaturated with respect to $BaSO_4$ that precipitation is favored. Organic matter may also mediate $BaSO_4$ formation; on suspended particles, Ba can be sorbed onto organic moieties that are later substituted by seawater sulfate (e.g., Martinez-Ruiz et al., 2019, 2020). Likewise, organic–mineral interfaces can drive heterogeneous $BaSO_4$ nucleation from undersaturated

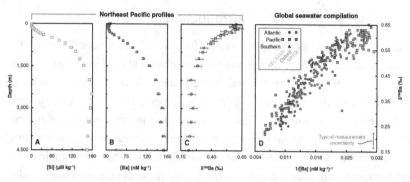

Figure 1 | Barium in seawater. Profiles of dissolved **A:** [Si], **B:** [Ba], and **C:** δ^{138}Ba (see text for definition) from the northeast Pacific Ocean at "SAFe" (30° N, 140°W; Geyman et al., 2019). **D:** Property–property plot showing a global compilation of δ^{138}Ba against 1/[Ba]. Colors correspond to laboratory: GEOMAR (Cao et al., 2020a; b), Oxford (Hsieh & Henderson, 2017; Bridgestock et al., 2018; Hemsing et al., 2018), and WHOI (Horner et al., 2015; Bates et al., 2017; Geyman et al., 2019). These data indicate excellent agreement between laboratories and similarity of the δ^{138}Ba–1/[Ba] relationship between – and within – the major ocean basins. The processes generating the gradient and spread of this array are described in Sections. 2.1 and 2.2.2.

solutions (Deng et al., 2019). Altogether, these processes render $BaSO_4$ precipitation possible, even in strongly undersaturated environments (e.g., Horner et al., 2017). Continued microbial action eventually destroys the microenvironment, exposing $BaSO_4$ precipitates to undersaturated seawater, where dissolution may begin. We refer to $BaSO_4$ formed via these processes as "pelagic," thereby distinguishing these precipitates from other types of marine $BaSO_4$ (e.g., hydrothermal, diagenetic; Hanor, 2000). Pelagic $BaSO_4$ are ubiquitous in marine particulate matter, which can contain up to 10^4 crystals per L (e.g., Dehairs et al., 1980). The formation and export of pelagic $BaSO_4$ connects the cycling of Ba with the marine biogeochemical processes of nutrient consumption, organic matter cycling, and carbon export. It thus follows that the isotope geochemistry of Ba – in seawater, sediments, and during Earth's past – may similarly offer a window into the same biogeochemical processes. The history of mass-dependent Ba-isotopic measurements dates back only ten years to a landmark study by von Allmen et al. (2010). Subsequent studies have investigated and identified significant Ba-isotopic variations in Earth's surficial envelopes, including seawater (e.g., Horner et al., 2015) and sediments (Bridgestock et al., 2018; Hemsing et al., 2018; Crockford et al., 2019a;

Box 1 | Marine barite saturation

The saturation state of a solution, such as seawater, with respect to barite (Ω_{barite}) is defined as the ratio of the Ba and sulfate IAP (ion activity product), normalized by the barite solubility product (K_{sp}):

$$\Omega_{\text{barite}} = (\{Ba^{2+}\} \cdot \{SO_4^{2-}\}) / K_{\text{sp}}$$

Values of $\Omega_{\text{barite}} < 1$ indicate undersaturation, $\Omega_{\text{barite}} = 1$ indicates perfect saturation, and $\Omega_{\text{barite}} > 1$ indicates supersaturation. Homogeneous nucleation of barite occurs above Ω_{barite} of 8 (Nancollas & Purdie, 1963).

Marine barite saturation is controlled by the physical and chemical properties of seawater (Church & Wolgemuth, 1972). The most important physical controls are temperature and pressure. Increasing temperature lowers Ω_{barite} by decreasing the Ba and sulfate IAP, and by increasing K_{sp}. Pressure increases both the IAP and K_{sp}, rendering opposing effects on Ω_{barite}. Since K_{sp} exhibits a stronger pressure dependence than the IAP, for a fixed Ba and sulfate molality, Ω_{barite} decreases with increasing water depth. Barium molality and salinity are the most important chemical controls over Ω_{barite}. Sulfate molality does not influence modern Ω_{barite} since sulfate is conservative with respect to salinity. However, secular increases in marine sulfate levels have likely driven a substantial decrease in marine [Ba] over Earth's history (e.g., Walker, 1983; Section 2.3.). Barium molality affects the Ba and sulfate IAP, such that increases in [Ba] drive higher Ω_{barite}. Higher salinities cause lower Ω_{barite}: Elevated salinity lowers IAP and increases K_{sp}, though the salinity effect on K_{sp} is minor over the range of salinities encountered in seawater.

Ω_{barite} at the sea surface

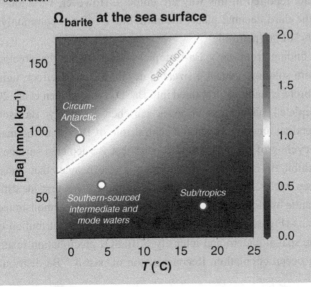

Spatial variations in the physical and chemical composition of seawater renders large variations in Ω_{barite} (Monnin et al., 1999; Rushdi et al., 2000). Saturation may be exceeded in the cold, Ba-rich waters in the surface of the Southern Ocean, whereas the hot, salty, Ba-poor subtropics exhibit some of the lowest Ω_{barite} in the open ocean (see the figure). No region of the modern ocean exceeds the threshold Ω_{barite} of 8 necessary for homogeneous $BaSO_4$ precipitation. The figure is based on 108 simulations performed across the parameter space, conducted using PHREEQC (Parkhurst & Appelo, 1999). Barite saturation was computed for a seawater-like thermodynamic system containing Cl, Na, Mg, SO_4^{2-}, Ca, K, HCO_3^-, Br, B, Sr, F, and Ba at a pH of 8.1. Simulation results are plotted using Ocean Data View (Schlitzer, R., https://odv.awi.de, 2018).

Geyman et al., 2019; Liu et al., 2019), all the way down to subducting slabs of lithosphere (Nielsen et al., 2020). The diversity of applications of Ba isotopes has led to a diversity of reporting conventions. In this Element, we report Ba-isotopic data in the delta (δ) notation as variations in the $^{138}Ba/^{134}Ba$ ratio relative to the National Institute of Standards and Technology (NIST) Standard Reference Material (SRM) 3104a standard, regardless of how data were originally reported (see Box 2):

$$\delta^{138}Ba = (^{138}Ba/^{134}Ba)_{sample} / (^{138}Ba/^{134}Ba)_{standard} - 1 \qquad (1)$$

Isotope data reported in this way are unitless. However, variations typically occur at the third decimal place, meaning that $\delta^{138}Ba$ is commonly reported with "units" of ‰ (per mille).

A key finding from early Ba-isotopic studies is that the precipitation of $BaSO_4$ from solution renders a sizeable negative isotope effect; light Ba isotopes are preferentially incorporated into $BaSO_4$ (von Allmen et al., 2010), and the Ba-depleted residual solution is enriched in heavy Ba isotopes by a corresponding amount. Given the central role of $BaSO_4$ in the marine geochemical cycling of Ba, one would predict that Ba-depleted surface seawater should exhibit Ba-isotopic compositions that are heavier than those of Ba-replete deep waters. Indeed, this is precisely the pattern first identified in the South Atlantic by Horner et al. (2015) and has since been demonstrated in other ocean basins (Figure 1). On a global basis, the marine distribution of $\delta^{138}Ba$ reflects the spatial pattern and intensity of $BaSO_4$ precipitation relative to the prevailing ocean circulation. Reconstruction of past $\delta^{138}Ba$ distributions can

Box 2 | Barium isotope notation and standardization

Barium possesses seven stable isotopes: ^{130}Ba (0.106%), ^{132}Ba (0.101%), ^{134}Ba (2.417%), ^{135}Ba (6.592%), ^{136}Ba (7.854%), ^{137}Ba (11.232%), and ^{138}Ba (71.698%; see the figure, reproduced from Dempster, 1936), with an atomic mass of 137.327 Da.

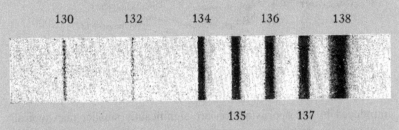

Reprinted figure with permission from Dempster, A. J., *Physical Review*, **49**, p. 947, 1936. Copyright 1936 by the American Physical Society.

Assuming isotope data are reported using a "heavy-over-light" convention, there are twenty-one possible permutations with which to report Ba stable isotopic data. In spite of myriad possibilities, most studies report Ba-isotopic data using either δ^{138}Ba or δ^{137}Ba notation, which represent variations in the ^{138}Ba/^{134}Ba or ^{137}Ba/^{134}Ba ratios, respectively. While both of these common notations represent the same underlying mass-dependent isotopic variations, conversion between notations can be confounded by the use of different reference materials. In this Element, we use NIST SRM 3104a as the δ^{13x}Ba $\equiv 0$ standard, and report all variations in the δ^{138}Ba notation.

Comparing Ba-isotopic data reported in ^{137}Ba/^{134}Ba notation with those reported in ^{138}Ba/^{134}Ba (and vice versa) requires recalculation based on an assumed mass fractionation "law" (e.g., Wombacher & Rehkämper, 2004). Assuming variations follow the kinetic law, data reported in δ^{137}Ba notation can be converted to δ^{138}Ba using:

$$\delta^{138}\text{Ba} = [(\delta^{137}\text{Ba} / 1{,}000)^{\beta} - 1] \times 1{,}000$$

Under the kinetic law, β is defined as:

$$\beta = \ln(m_{138} / m_{134}) / \ln(m_{137} / m_{134})$$

where m is the exact mass of the Ba isotope 138, 137, or 134. Given current analytical precision, the following approximation is generally valid:

$$\delta^{138}\text{Ba} \approx \delta^{137}\text{Ba} \times \beta$$

Prior to 2017, a number of studies reported Ba-isotopic data relative to "in-house" reference materials, rather than NIST SRM 3104a. Barium-isotopic data for these samples can be renormalized to NIST SRM 3104a using the following conversion:

$$\delta^{138}Ba_{spl.-NIST} = \delta^{138}Ba_{spl.-in\ house} + \delta^{138}Ba_{in\ house-NIST}$$
$$+ (\delta^{138}Ba_{spl.-in\ house} \times \delta^{138}Ba_{in\ house-NIST}) / 1,000$$

which can be approximated as:

$$\delta^{138}Ba_{spl.-NIST} \approx \delta^{138}Ba_{spl.-in\ house} + \delta^{138}Ba_{in\ house-NIST}$$

The simpler conversion is similarly valid since any systematic errors introduced by the approximation are significantly smaller than typical analytical uncertainty. To avoid such errors, however, we recommend that future studies report Ba-isotopic variations based on the $^{138}Ba/^{134}Ba$ ratio and relative to NIST SRM 3104a, in place of – or in addition to – authors' preferred notation.

thus reveal key features of ancient marine biogeochemistry. Likewise, the isotopic composition of Ba in resultant $BaSO_4$ (hereafter $\delta^{138}Ba_{barite}$) will strongly depend on the quantity and isotopic composition of Ba in the starting fluid reservoir. Since different reservoirs contain different quantities of Ba and exhibit characteristic Ba isotope compositions, $\delta^{138}Ba_{barite}$ can provide diagnostic information on the sources and cycling of Ba in the environment in which those $BaSO_4$ formed.

2 Underpinnings

What controls $\delta^{138}Ba$? Assuming that $BaSO_4$ cycling has been the principal dissolved–particulate transformation of Ba occurring in the ocean over much of Earth's history, there are three main controls: $\Delta^{138}Ba_{barite-dBa}$, the intrinsic Ba-isotopic fractionation factor between product $BaSO_4$ and reactant dissolved Ba (dBa; Section 2.1); $\delta^{138}Ba_{dBa}$, the Ba-isotopic composition of the dissolved Ba source (Section 2.2); and, time, in that geological processes likely drove secular changes in deep $\delta^{138}Ba_{dBa}$ (Section 2.3). These controls apply to the interpretation of any sedimentary record of past $\delta^{138}Ba_{dBa}$, $BaSO_4$ or not.

2.1 Constraints and Controls on $\Delta^{138}Ba_{barite-dBa}$

The isotopic fractionation factor between the product $BaSO_4$ and reactant dissolved Ba, is defined as:

$$\Delta^{138}Ba_{barite-dBa} \equiv \delta^{138}Ba_{barite} - \delta^{138}Ba_{dBa} \tag{2}$$

This definition of $\Delta^{138}Ba_{barite-dBa}$ can be used to estimate $\alpha_{barite-dBa}$:

$$\alpha_{barite-dBa} = {}^{138/134}Ba_{barite} / {}^{138/134}Ba_{dBa}$$

using the approximation: (3)

$$\Delta^{138}Ba_{barite-dBa} \approx 1,000 \times \ln(\alpha_{barite-dBa}) \tag{4}$$

To date, there are only two studies that have experimentally investigated $\Delta^{138}Ba_{barite-dBa}$. The initial study by von Allmen (2010) found a mean $\Delta^{138}Ba_{barite-dBa} = -0.32 \pm 0.03‰$ based on the results of two experiments conducted at 21 and 80°C. More recently, Böttcher et al. (2018) observed $\Delta^{138}Ba_{barite-dBa} = -0.25‰$ for $BaSO_4$ precipitation in the presence of methanol, and between -0.28 and $-0.35‰$ for $BaSO_4$ formation during transformation from gypsum ($CaSO_4 \cdot 2H_2O$) conducted between 4 and 60°C. There appears to be no strong temperature dependence on experimentally determined $\Delta^{138}Ba_{barite-dBa}$, at least over the ranges tested thus far. Salinity (i.e., ionic competition; Horner et al., 2011), precipitation rate (e.g., Mavromatis et al., 2020), and pressure (e.g., Geyman et al., 2019) are plausible additional controls on $\Delta^{138}Ba_{barite-dBa}$ that have yet to be systematically investigated (see Section 5).

In contrast to laboratory experiments, there are three independent lines of evidence suggesting $\Delta^{138}Ba_{barite-dBa}$ is larger than $\approx -0.3‰$ in the marine environment, and likely closer to $\approx -0.5‰$. First, by assuming open system fractionation at steady state, one can estimate $\Delta^{138}Ba_{barite-dBa}$ by regressing individual depth profiles of dissolved $\delta^{138}Ba$ against [Ba] and assume that the trend is driven by $BaSO_4$ precipitation. This approach yields values ranging from $-0.39\pm0.04‰$ for the South Atlantic (Horner et al., 2015) to $-0.45 \pm 0.08‰$ for the North Atlantic (Bates et al., 2017). Rather than regressing individual depth profiles, Hsieh & Henderson (2017) regressed the compilation of global seawater data using the Southern Ocean as a common starting point to yield $\Delta^{138}Ba_{barite-dBa} = -0.58 \pm 0.10‰$. While instructive, these regression models are oceanographically questionable. The models assume that marine $BaSO_4$ precipitation occurs within a non-dimensional reactor, thereby neglecting important oceanographic processes such as physical transport. As such, estimates of $\Delta^{138}Ba_{barite-dBa}$ derived by regression of seawater data will tend to underestimate the true fractionation factor, and can only provide a regional average that may mask local variations.

The second line of evidence that $\Delta^{138}Ba_{barite-dBa}$ is closer to $-0.5‰$ comes from sediments. Bridgestock et al. (2018) examined $\delta^{138}Ba$ in total digests of six core-top and recent down-core sediments from the South Atlantic. They observed that the patterns of sedimentary $\delta^{138}Ba$ were best explained as reflecting a mixture between Ba hosted in terrigenous material (possessing $\delta^{138}Ba$ between -0.1 and

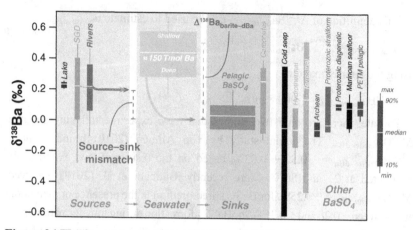

Figure 2 | The ins and outs of the barium cycle. Data have been grouped to illustrate isotopic differences (or similarities) between major surficial Ba reservoirs. From left: major Ba sources (SGD, submarine groundwater discharge; Section 2.2.1), dissolved Ba in seawater (Section 2.2.2), the principal Ba sinks (Section 2.1), and other $BaSO_4$ representing a range of depositional environments and ages (PETM, Paleocene–Eocene Thermal Maximum; Section 4). Data are displayed as per the legend on the right. This compilation illustrates the implied fractionation associated with $BaSO_4$ precipitation, $\Delta^{138}Ba_{barite-dBa}$ (Section 2.1), as well as the modern source–sink imbalance (Section 2.2.1). Data sources are described in the relevant sections.

0.0‰; Nan et al., 2018) and authigenic Ba ($\delta^{138}Ba = +0.09 \pm 0.04$‰, $n = 68$; Figure 2). Assuming that the authigenic Ba was derived from dissolved Ba in seawater above 500 m, Bridgestock et al., (2018) concluded that $\Delta^{138}Ba_{barite-dBa}$ must be between -0.4 and -0.5‰. More recently, Crockford et al. (2019a) measured $\delta^{138}Ba$ in $BaSO_4$ isolates from 61 core-top sediments. These samples exhibited a mean $\delta^{138}Ba = +0.04 \pm 0.06$‰ (Figure 2), again implying an average $\Delta^{138}Ba_{barite-dBa} = -0.5 \pm 0.1$‰.

Third, data for in situ-collected particles also imply a larger $\Delta^{138}Ba_{barite-dBa}$. Horner et al. (2017) calculated $\Delta^{138}Ba_{barite-dBa}$ as -0.41 ± 0.09‰ ($n = 20$) based on the average Ba isotope offset between the particulate Ba excess (i.e., above lithogenic background) and total dissolvable Ba in Lake Superior. These authors also examined a depth profile of marine particles collected nearby to the dissolved $\delta^{138}Ba$ profile presented in Horner et al. (2015) and found $\Delta^{138}Ba_{barite-dBa} = -0.53 \pm 0.04$‰ ($n = 6$). Most recently, Cao et al., (2020a) reported $\Delta^{138}Ba_{barite-dBa} = -0.46 \pm 0.11$‰ ($n = 19$) for samples collected above 150m in the South China Sea.

. Existing data do not allow a confident assessment of whether there are systematic regional or vertical variations in $\Delta^{138}Ba_{barite-dBa}$ in seawater. Regardless, there is a clear mismatch between experiment- and field-based estimates of $\Delta^{138}Ba_{barite-dBa}$. The latter are considerably larger, implying the existence of at least one additional process operating in the marine realm that is not captured by existing laboratory studies. Reconciling this mismatch should be considered a priority for future studies (see Section 5).

2.2 Controls on $\delta^{138}Ba_{dBa}$

In this section, we discuss the processes that influence $\delta^{138}Ba_{dBa}$. This discussion is subdivided into global (Section 2.2.1), regional (Section 2.2.2), and local (Section 2.2.3) processes.

2.2.1 Barium Budget of the Oceans

The main source of Ba to the modern ocean is thought to derive from the weathering of crustal silicates, followed by transport and delivery to seawater by rivers and groundwater (e.g., Paytan & Kastner, 1996). The upper continental crust possesses high Ba contents of ≈ 600 μg g^{-1} (e.g., Wedepohl, 1995) and $\delta^{138}Ba = 0.00 \pm 0.04‰$ (Nan et al., 2018). Mantle-derived silicates, such as mid-ocean ridge basalt and, by inference, the mantle itself, possess significantly lower Ba contents (~10s to 100s of μg g^{-1}) and $\delta^{138}Ba$ that is similar to – or slightly heavier than – the upper continental crust (c.f. Nielsen et al., 2018; Li et al., 2020). The origin of the low solid Earth Ba content and, to a lesser extent its variation in $\delta^{138}Ba$, relates to Ba being both fluid mobile and highly incompatible during partial melting. Indeed, there is significant and ongoing interest in high-temperature applications of Ba isotopes to study the sources and cycling of material between the major lithospheric reservoirs (e.g., Guo et al., 2020; Nielsen et al., 2020).

Biological and chemical weathering can modulate the Ba-isotopic composition of Ba delivered to the ocean. Bullen and Chadwick (2016) showed that plant "biolifting" preferentially removes isotopically light Ba from soils. Chemical weathering of soils is more complex; the isotopic composition of Ba liberated to solution reflects an interplay between dissolution, precipitation, and adsorption, whereby each process may possess a unique fractionation factor (e.g., Charbonnier et al., 2020; Gong et al. 2020). These interacting processes are likely responsible for the significant inter- (Cao et al., 2020b) and intra-riverine (Charbonnier et al., 2020; Gou et al., 2020) variability in $\delta^{138}Ba$, with dissolved Ba loads exhibiting isotopic compositions from ≈ 0.0 to $+ 0.5‰$. Although a large range, most riverine samples exhibit dissolved $\delta^{138}Ba$ that is heavier than the upper continental crust by about +0.2‰. Extrapolated globally, the data of Cao et al. (2020b), Charbonnier et al. (2020), and Gou et al. (2020) imply a discharge- and [Ba]-weighted global riverine flux of $\delta^{138}Ba \approx +0.2$‰, slightly heavier than the mean for groundwater discharge ($\approx +0.1$‰; Mayfield et al., 2021).

In the modern ocean, $BaSO_4$ sedimentation constitutes the main output of Ba from seawater (e.g., Paytan & Kastner, 1996). Core-top studies indicate that this output term possesses, on average, $\delta^{138}Ba$ between 0.0 and +0.1‰ (Bridgestock et al., 2018; Crockford et al., 2019a). Thus, the following equality describes the global mass balance of Ba isotopes at steady state:

$$\sum_{i=1}^{n} (f_{input} \cdot \delta_{input})_i = \delta_{barite} \qquad (5)$$

where f is the relative magnitude and δ is the Ba isotope composition of flux term i. Assuming that the modern Ba cycle is at steady state, Equation 5 demands that the n input fluxes sum to an average $\delta^{138}Ba$ between 0.0 and +0.1‰. Based on existing riverine and groundwater data described above, this is not the case; known Ba inputs are too heavy to close the isotopic budget of the ocean. The isotopic composition of the "missing" Ba needed to balance the budget depends on the magnitude of the uncharacterized term(s) – the more minor the flux magnitude, the lighter the composition it must possess. That we cannot more precisely define $\delta^{138}Ba$ for the missing term(s) is because the flux magnitudes are also highly uncertain. For example, the recent compilation by Mayfield et al. (2021) suggests that riverine and groundwater Ba inputs are between $\approx$7–10 Gmol yr^{-1}, with groundwater accounting for 6–36% of the combined flux. If hydrothermal (e.g., Dickens et al., 2003), cold seep (e.g., Torres et al., 2003), or estuarine (e.g., Li & Chan, 1979) Ba fluxes are comparable, total Ba delivery to the ocean could approach 22 Gmol yr^{-1}. Given the total marine Ba inventory of $\approx$150 Tmol, an input flux range of 7–22 Gmol yr^{-1} implies a Ba residence time between 7–21 kyr. Identifying and characterizing putative Ba sources is needed to narrow these flux estimates, which will greatly improve the utility of Ba-based proxies in paleoceanography (see Section 5).

2.2.2 Role of Ocean Circulation

On a regional basis, one of the most important controls on $\delta^{138}Ba_{dBa}$ is ocean circulation – its overall strength and geometry. The strength of ocean circulation, relative to the strength of the vertical cycle of Ba (i.e., precipitation and regeneration of $BaSO_4$), sets the gradient and spread in the modern array of dissolved $\delta^{138}Ba$–1/[Ba] (Figure 1). Broadly speaking, this is because $BaSO_4$ precipitation and regeneration are vertically segregated. Dissolved Ba removal and fractionation of Ba isotopes occur (mostly) at or near the surface. In contrast, $BaSO_4$ dissolution can occur throughout the water column. Thus, the longer a water mass remains out of contact with the surface, the more isotopically light Ba can accumulate through regeneration of $BaSO_4$. Indeed, old and deep water masses with significant concentrations of regenerated nutrients and low $[O_2]$, such as Pacific Deep Water, possess some of the highest $[Ba] \approx 150$ nmol kg^{-1} and lightest $\delta^{138}Ba$ ($\approx +0.2‰$) in the global ocean (Geyman et al., 2019). Young surface water masses with high $[O_2]$ and low nutrient contents exhibit low $[Ba] \approx 35$ nmol kg^{-1} and heavy $\delta^{138}Ba \approx +0.6‰$. Almost all open ocean seawater samples fall on a roughly linear array between these two extremes in $\delta^{138}Ba$–1/[Ba] space (Figure 1). The spread and gradient of this array is an emergent property of the modern Ba cycle; stronger overturning circulation – or a weaker vertical Ba cycle – will diminish the contrast between the extrema and steepen the global $\delta^{138}Ba$–1/[Ba] array. Conversely, weaker overturning circulation (or a stronger vertical Ba cycle) will enhance the spread and yield a gentler slope of $\delta^{138}Ba$–1/[Ba]. Over short timescales, such as glacial–interglacial cycles, the strength of the vertical Ba cycle and ocean overturning circulation are likely positively correlated, although this may not be true over geological timescales (see Section 2.3).

The geometry of ocean circulation can similarly influence $\delta^{138}Ba_{dBa}$ and thus the Ba-isotopic composition of resultant $BaSO_4$. The physical processes of upwelling, deep water formation, and mixing within the ocean interior are spatially segregated, thereby driving characteristic variations in the Ba content and Ba-isotopic composition of the major water masses. To illustrate this point, we draw on a simplified version of the modern overturning circulation considering only the Atlantic Ocean (Figure 3). This simplification is necessary given that there are currently no Ba-isotopic data for the Indian Ocean and too few data for the Pacific to confidently constrain global meridional transports. In major upwelling areas, such as the Southern Ocean, Circumpolar Deep Water (CDW) is ventilated, bringing high $[Ba]$ (≈ 90 nmol kg^{-1}) and light $\delta^{138}Ba$ between +0.3 to +0.4‰ to the surface (Hsieh & Henderson, 2017). Resultant $BaSO_4$ precipitates are predicted to exhibit $\delta^{138}Ba_{barite}$ that is $\approx -0.5‰$ lighter than seawater (i.e., $\approx -0.2‰$), which was observed by Crockford et al. (2019a) in a single core-top sample from the Pacific Sector of the Southern Ocean. Waters that become entrained in the low-productivity southern circuit experience minimal Ba drawdown before eventually subducting and becoming AABW. During subduction, waters in contact with the Antarctic continent may accumulate additional Ba from sedimentary sources (e.g., Hoppema et al., 2010). Given that end-member AABW (+0.26‰; Bates et al., 2017) is $\approx 0.1‰$ lighter than upwelled CDW (+0.3 to +0.4‰; Hsieh & Henderson,

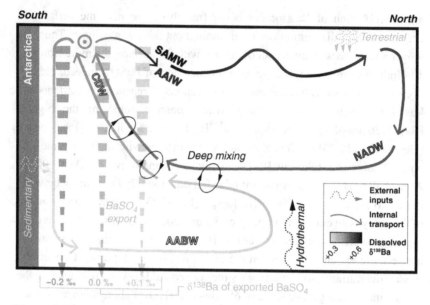

Figure 3 | The barium cycle in the Atlantic Ocean. The Atlantic overturning is dominated by two 'circuits' (Talley, 2013) – a northern circuit that is biologically productive, and a southern circuit, which is not. Differences in productivity (and attendant $BaSO_4$ export) render distinct differences in downstream $\delta^{138}Ba_{dBa}$, indicated by the color shading. Waters from both circuits are mixed in the deep Atlantic and ventilated in the Southern Ocean, thus closing the circuits. The arrow pointing toward the reader indicates strong zonal flow associated with the Antarctic Circumpolar Current. The expected Ba-isotopic composition of pelagic $BaSO_4$ exported to the seafloor (dashed arrows) is shown at the bottom, highlighting the importance of both circulation *and* biological productivity on resultant $\delta^{138}BaSO_4$. Dotted arrows indicate sources of "new" Ba to the ocean, which may locally influence $\delta^{138}Ba_{dBa}$

2017), it is likely that benthic Ba sources close to Antarctica possess isotopically light $\delta^{138}Ba$ (Figure 3). Regardless of this source, waters in the southern circuit are characterized by elevated [Ba] (≈ 100 nmol kg^{-1}) and light $\delta^{138}Ba$ between +0.2 and +0.3‰ (Bates et al., 2017; Hemsing et al., 2018). In contrast, waters that enter the high-productivity northern circuit undergo extensive modification during ventilation, presumably due to Ba removal into $BaSO_4$. As waters transit northward, surface dissolved [Ba] decreases from ≈ 90 to ≈ 50 nmol kg^{-1} (e.g., Jacquet et al., 2007), concomitant with an increase in $\delta^{138}Ba$ from $\approx +0.3$ to +0.5‰. Pelagic $BaSO_4$ derived from these waters are expected to reflect the spatial pattern seen in dissolved $\delta^{138}Ba$ (Figure 3). These Ba-depleted waters are eventually subducted and entrained in Subantarctic Mode Water

(SAMW) and Antarctic Intermediate Water (AAIW), where they spread north-ward, feeding Ba-poor and heavy δ^{138}Ba into low-latitude thermoclines. In the gyres north of 40°S, surface waters exhibit low [Ba] (< 40 nmol kg^{-1}) and heavy δ^{138}Ba$_{dBa} \approx +0.6$‰ (Horner et al., 2015). Resultant pelagic (Horner et al., 2017) and core-top (Bridgestock et al., 2018; Crockford et al., 2019a) BaSO$_4$ pos-sesses δ^{138}Ba $\approx +0.1$‰ (Figure 3). In the north Atlantic, near-surface waters are cooled (Talley, 2013). Strong winter convection leads to the formation of North Atlantic Deep Water (NADW), carrying surface-like [Ba] and δ^{138}Ba (≈ 50 nmol kg^{-1}; $\approx +0.4$ to $+0.5$‰; Bates et al., 2017) down into the ocean interior and back towards the Southern Ocean. Additional isotopically heavy Ba may be added to deep waters by hydrothermalism (Hsieh et al., 2021). During southward transport, NADW mixes with AABW, eventually forming CDW. Subsequent upwelling of CDW in the Southern Ocean closes the circuit (Figure 3).

Since BaSO$_4$ precipitation is largely confined to the upper layers of the ocean, the range of δ^{138}Ba observed in surface seawater ($\approx +0.3$ and $+0.6$‰) dictates the total range of δ^{138}Ba in pelagic precipitates (-0.2 to $+0.1$‰), assuming a constant Δ^{138}Ba$_{barite-dBa} \approx -0.5$‰. On a globally integrated basis, the δ^{138}Ba of pelagic BaSO$_4$ must sum to the mean oceanic input value, presumed to be between 0.0 and $+0.1$‰ (Section 2.1.1). However, at any one location, the δ^{138}Ba$_{barite}$ of locally formed precipitates will depend on the isotopic composition of the local Ba source, which depends on the geometry and strength of ocean circulation (relative to the vertical cycle of Ba). For example, local δ^{138}Ba$_{dBa}$ is largely dictated by the amount of BaSO$_4$ export that occurred "upstream" of that locality. Likewise, any changes in local BaSO$_4$ export will primarily manifest in downstream δ^{138}Ba$_{dBa}$ and resultant δ^{138}Ba$_{barite}$ (Figure 3). Thus, it is not possible to unambiguously interpret a single down-core record of δ^{138}Ba$_{barite}$ in terms of either the geometry or relative strength of ocean circulation without additional constraints. Such constraints could take the form of independent circulation estimates (e.g., from other geochemical proxies, models), or, ideally, by obtaining multiple spatially distributed, contemporaneous records of δ^{138}Ba$_{barite}$. We do not prescribe a set number of records that might be needed – this will depend on the time period and perturbation being studied – only that researchers consider this important spatial dependency when interpreting the proxy. Rather than being a weakness however, this is arguably one of the key strengths of the δ^{138}Ba$_{barite}$ proxy, especially when combined with BaSO$_4$ mass accumulation rates, which are sensitive to local carbon export (Eagle et al., 2003); together, these two Ba-based proxies can be used to distinguish between local and regional changes in export and can offer novel constraints on the geometry of ocean circulation at key transitions in Earth's history (see Section 4).

2.2.3 Local Influences

We identify three further processes that may influence $\delta^{138}Ba_{dBa}$ (and thus resultant $\delta^{138}Ba_{barite}$) on a local scale: pelagic microenvironments, intra-marine Ba sources, and Ba in non-marine environments.

The microenvironment-mediated model of precipitation is the most widely accepted solution to the $BaSO_4$ paradox. However, this model implies that local processes could exert some influence on $\delta^{138}Ba_{dBa}$. The starting composition for a microenvironment is likely similar to ambient seawater and with $\Omega_{barite} < 1$. For a microenvironment to become favorable to $BaSO_4$ precipitation, it must develop [Ba] in excess of ambient seawater. An obvious source of dissolved Ba ions to pelagic microenvironments is from the respiration of organic matter. Isotope tracer experiments performed by Ganeshram et al. (2003) suggest that the fraction of respired Ba in resultant $BaSO_4$ is at least 50%. This fraction is likely even greater in strongly $BaSO_4$-undersaturated environments (Horner et al., 2017). Thus, dissolved Ba within microenvironments – and by extension, resultant pelagic $BaSO_4$ – should reflect a mixture of Ba ions derived from ambient seawater and respired organic matter. This possibility was suggested by Horner et al. (2015), and is one putative mechanism for the discrepancy between lab- and field-based estimates of $\Delta^{138}Ba_{barite-dBa}$. For example, if Ba associated with organic matter were fractionated with respect to ambient seawater by -0.3 to $-0.5‰$, and respired organic matter contributed half of the dissolved Ba available for precipitation (with seawater contributing the remainder), resultant pelagic $BaSO_4$ could exhibit $\delta^{138}Ba \approx 0.5‰$ lighter than ambient seawater, even if $\Delta^{138}Ba_{barite-dBa}$ was only -0.25 to $-0.35‰$ (see Section 2.1). Several other factors may be important, such as: the type of substrate organic matter, which may exhibit regional differences in both Ba content and Ba-isotopic composition; the geometry and permeability of microenvironments, likely also depending on the types of particle; and, the process of organic matter respiration that could further fractionate dissolved Ba isotopes. All are presently poorly constrained and require additional characterization (see Section 5).

In addition to microenvironments, intra-marine Ba sources may also influence $\delta^{138}Ba_{dBa}$. In seawater, dissolved $\delta^{138}Ba$ generally decreases with depth, meaning that $\delta^{138}Ba_{dBa}$ could vary over the depth ranges where $BaSO_4$ is forming. The question of where $BaSO_4$ is forming is difficult to definitively answer. The peak in particulate Ba concentrations is generally at the top of the mesopelagic layer (200 m), decreasing monotonically thereafter (e.g., Dehairs et al., 1980; Bishop, 1988). Together with dissolved profile data, these observations imply that the vast majority of dissolved Ba removal occurs at or above 200 m, where variations in $\delta^{138}Ba_{dBa}$ are generally muted. However, evidence from radium isotopes suggests that some $BaSO_4$ forms deeper, perhaps even throughout the entire water column (e.g., van Beek et al., 2007), in which case deeper forming $BaSO_4$ may exhibit lighter-than-expected $\delta^{138}Ba_{barite}$. Local point sources may also influence $\delta^{138}Ba_{dBa}$. Rivers have long been recognized to increase coastal [Ba], and data from Hsieh and Henderson (2017) and Cao et al. (2020b) indicate that riverine inputs lower near-shore $\delta^{138}Ba_{dBa}$ in the South Atlantic and East China Sea, respectively. Likewise, benthic Ba inputs – from hydrothermalism, cold seeps, and diagenetic remobilization within margin sediments – can release significant quantities of Ba to deep waters. If saturation with respect to $BaSO_4$ is exceeded, these Ba sources can induce benthic $BaSO_4$ precipitation (e.g., Paytan et al., 2002; Torres et al., 2003), likely modifying $\delta^{138}Ba_{dBa}$ of the source to values distinct from background seawater (e.g., Hsieh et al., 2021). Aside from hydrothermalism, these intra-marine benthic Ba sources await Ba-isotopic characterization (see Section 5).

Non-marine environments (i.e., terrestrial settings) are expected to exhibit a wide range of $\delta^{138}Ba_{dBa}$. This expectation is based on the wide range of $\delta^{138}Ba$ observed in non-marine $BaSO_4$, which vary from roughly -0.6 to $+0.6‰$ (Crockford et al., 2019a). A recent study by Tieman et al. (2020) provides evidence for this prediction; these authors analyzed produced water from a shale gas play implying that deep brines possess $\delta^{138}Ba_{dBa}$ up to $+1.5‰$. Why non-marine Ba sources exhibit such wide ranges will depend on a number of geological factors that are likely specific to that terrane (lithology, prior precipitation, etc.; Hanor, 2000). These processes are insufficiently constrained for us to confidently discern any patterns. It is nonetheless probable that some of this variation derives from leaching and transport of Ba by fluids (e.g., van Zuilen et al., 2016a; Gong et al., 2020), which must remain segregated from fluids containing sulfate. Non-marine $BaSO_4$ will tend to form where Ba- and sulfate-rich fluids mix (e.g., Hanor, 2000). That non-marine Ba sources can possess such wide ranges in $\delta^{138}Ba_{dBa}$ is itself a valuable finding that underpins the second major use of $\delta^{138}Ba_{barite}$: to diagnose the origin of enigmatic sulfates in the geological record (Figure 2; Section 4). If $BaSO_4$ sedimentation constitutes the primary marine Ba output flux, marine $BaSO_4$ should, on average, exhibit a narrow range of $\delta^{138}Ba_{barite}$ between 0.0 and $+0.1‰$ (Bridgestock et al., 2018; Crockford et al., 2019a). In contrast, $BaSO_4$ derived from non-marine sources can exhibit values far outside this range, a signature that we contend is diagnostic of non-marine environments. Environments that become semi-restricted may exhibit a progressive distillation of $\delta^{138}Ba$, analogous to calcium isotopes (e.g., Blättler & Higgins, 2014). As with calcium, interpretations derived solely from $\delta^{138}Ba_{barite}$ may be ambiguous – non-marine $BaSO_4$ could conceivably exhibit marine-like $\delta^{138}Ba_{barite}$ – underscoring the importance of considering the regional geological context in any interpretations.

2.3 Geological Influences on δ^{138}Ba

We now consider two factors that do not appreciably influence δ^{138}Ba$_{dBa}$ in the modern ocean, but were likely important in Earth's past – the concentration of sulfate in seawater and the total productivity of the biosphere. (A putative third factor, non-steady state behavior arising from external perturbations, may also have been important at certain times, as discussed by Dickens et al., 2003.) Multiple lines of evidence suggest that both marine [SO$_4^{2-}$] and productivity have increased over Earth's history (e.g., Walker, 1983; Crockford et al., 2018a, 2019b). Since the influence of these two processes is not easily tested in the field, we quantified their impact using a numerical model (Figure 4). This model is based on the two-box model developed by Dickens et al. (2003), with the addition of a Ba-isotopic mass balance. We used a fixed Δ^{138}Ba$_{barite-dBa}$ of -0.5‰ and assumed no Ba-isotopic fractionation during BaSO$_4$ dissolution. To further simplify the model, we neglected benthic Ba sources and forced surface Ba inputs to balance benthic outputs at δ^{138}Ba $= +0.1$‰, using a prescribed BaSO$_4$ burial flux of 18 Gmol yr^{-1} (Figure 4A). The flux of pelagic BaSO$_4$ in our model is related to both organic carbon export and ambient [Ba], as in Dickens et al. (2003). Carter et al. (2020) recently proposed a refinement of this relationship that eliminates the dependency on ambient [Ba]. We opted to retain this dependency given that, on geological timescales, a coupling between BaSO$_4$ fluxes and [Ba] is necessary to maintain steady state in the Ba cycle. The model has been run to steady state in the scenarios described below.

The concentration of sulfate in seawater sets the total Ba-carrying capacity of the ocean by controlling mean ocean Ω_{barite}, which, on a global basis, is slightly below unity (Monnin et al., 1999; Box 1). Thus, when seawater possessed less sulfate, [Ba] was likely higher, maintaining average Ω_{barite} at values similar to modern (Walker, 1983). Indeed, in the deep anoxic and sulfidic Black Sea, where dissolved [SO$_4^{2-}$] is $\approx 60\%$ of open marine values, dissolved [Ba] is ≈ 500 nmol kg^{-1} and resultant $\Omega_{barite} > 1$ (Falkner et al., 1993). We parameterized variable marine [SO$_4^{2-}$] using [Ba]$_{sat.}$, the concentration of dissolved Ba when $\Omega_{barite} = 1$. Since most BaSO$_4$ dissolution occurs on the seafloor (McManus et al., 1994; Paytan & Kastner, 1996), the value of [Ba]$_{sat.}$ was set to that observed in porewaters from deep-sea sediments (≈ 0.3 µmol kg^{-1}; Paytan & Kastner, 1996). We then explored how changes in [Ba]$_{sat.}$ influenced the Ba inventory and isotopic composition of seawater over a range spanning 0.33–330 µmol kg^{-1} (Figure 4). The model results show that at higher [Ba]$_{sat.}$ (i.e., lower [SO$_4^{2-}$]), the marine Ba inventory is greatly increased, with the most pronounced increases observed in the surface ocean reservoir (relative to modern). This behavior arises from a greater regeneration of pelagic BaSO$_4$ fluxes in the deep ocean at higher

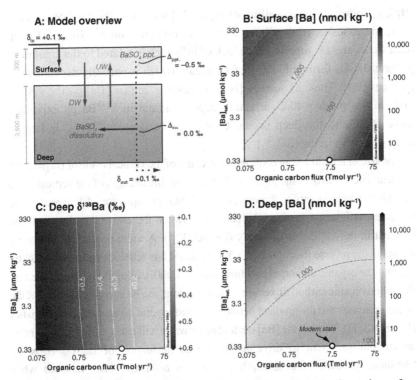

Figure 4 | Budget and balance of barium in the ocean. A: Overview of a steady-state two-box model used in simulations shown in panels **B–D**. The model is based on that of Dickens et al. (2003) with the addition of an isotopic component (this study). Solid arrows denote dissolved Ba inputs: an external source and upwelling (UW) of deep-water Ba to the surface ocean, barite dissolution and downwelling (DW) to the deep ocean. The dotted arrow denotes the flux of particulate Ba ($BaSO_4$), the outward flux of which balances external Ba inputs. Barite precipitation occurs with an isotopic effect, $\Delta^{138}Ba_{barite-dBa}$, of $-0.5‰$, whereas $BaSO_4$ dissolution occurs with no fractionation (i.e., $\Delta^{138}Ba_{dBa-barite} = 0.0‰$). Panels **B–D** show the response of **B:** surface ocean [Ba], **C:** deep-ocean $\delta^{138}Ba$, and **D:** deep-ocean [Ba] across a range of $[Ba]_{sat.}$ and organic carbon fluxes. Circles denote modern conditions. Surface $\delta^{138}Ba$ is $+0.6‰$ in all scenarios. While dissolved [Ba] in the surface and deep ocean are sensitive to both $[Ba]_{sat.}$ and the flux of organic carbon, the Ba-isotopic composition of the deep ocean depends almost exclusively on the size of the marine biosphere relative to the rate of ocean overturning. Data in panels **B–D** are based on the results of forty-nine simulations conducted across the parameter space and visualized using Ocean Data View (Schlitzer, R., https://odv.awi.de, 2018)

$[Ba]_{sat.}$, followed by upwelling of high-$[Ba]$ waters to the surface. In contrast, increases in $[Ba]_{sat.}$ have almost no influence on $\delta^{138}Ba_{dBa}$ of deep ocean seawater (Figure 4C). The lack of influence is related to the amount of pelagic $BaSO_4$ formed in the surface ocean, which is partially dependent on ambient $[Ba]$. Thus, as surface $[Ba]$ increases at higher $[Ba]_{sat.}$, the system maintains a modern-like surface-to-depth gradient in $\delta^{138}Ba$ by exporting more $BaSO_4$ to the deep ocean. Consequently, deep $\delta^{138}Ba_{dBa}$ is largely insensitive to marine $[SO_4^{2-}]$.

The size of the biosphere controls the strength of the vertical cycle of Ba (relative to ocean circulation). In the model, as in the ocean, the strength of the vertical cycle of Ba is tied to productivity through pelagic $BaSO_4$ precipitation. A less productive biosphere will tend to reduce the contrast between the extrema and steepen the gradient in the $\delta^{138}Ba-1/[Ba]$ array (Figures 1, 4). A more productive biosphere results in the reverse (Section 2.2.2). We explored this relationship in the model by varying the organic carbon export flux from 1–1,000% of modern values. We find that when organic carbon fluxes are ~1% of modern, the Ba carrying capacity of the ocean increases and the gradient between the surface and deep ocean $[Ba]$ vanishes (Figure 4B, D). Though the $[Ba]$ content of seawater will exceed $[Ba]_{sat.}$, it is not by an amount sufficient to drive homogeneous nucleation of $BaSO_4$ (Box 1). Significantly, the surface-to-depth gradient in $\delta^{138}Ba_{dBa}$ also vanishes when organic carbon export fluxes are low. When the vertical cycle is weak, the Ba-isotopic composition of the ocean becomes uniform at +0.6‰, and is controlled entirely by the isotopic contrast between the major Ba sources ($\delta^{138}Ba = +0.1‰$) and sinks ($\Delta^{138}Ba_{barite-dBa} = -0.5‰$). Such conservative behavior is analogous to that seen today for the stable isotope behaviors of other alkaline earth metals that possess long oceanic residence times – magnesium, calcium, and strontium (see respective Elements in this series).

This exercise highlights how secular increases in marine $[SO_4^{2-}]$ and organic carbon export over Earth's history have almost certainly driven a concomitant decrease in the concentration and residence time of Ba in the ocean. Importantly however, marine $[SO_4^{2-}]$ does not appreciably influence the surface-to-depth gradient in $\delta^{138}Ba_{dBa}$; this gradient depends only on organic carbon export, which controls the strength of the vertical Ba cycle relative to ocean circulation.

We highlighted a number of global, regional, local, and geological controls that influence $\delta^{138}Ba_{dBa}$ and, therefore $\delta^{138}Ba_{barite}$. Although these controls are reviewed primarily from the perspective of pelagic $BaSO_4$, these underpinnings can be used to interpret records of $\delta^{138}Ba_{dBa}$ recovered from any sedimentary archive (e.g., carbonates; Section 5). While this litany of controls may seem a significant barrier to confident interpretation of geological $\delta^{138}Ba$, the premise of the proxy is robust and bears restating: $\delta^{138}Ba_{barite}$ depends on the quantity and isotopic composition of Ba in the starting reservoir. Since different Ba reservoirs contain different quantities of Ba and often exhibit distinct $\delta^{138}Ba_{dBa}$, the utility of this tracer is that it can provide diagnostic information on the sources and cycling of Ba in the environment in which $BaSO_4$ formed.

3 Materials and Methods

3.1 Archives

Barite is an appealing archive of $\delta^{138}Ba_{dBa}$ for several reasons: $BaSO_4$ deposits are present throughout the geological record (e.g., Hanor, 2000), their preservation in deep-sea sediments exceeds that of many other co-occurring phases (e.g., organic carbon; biogenic silica; Paytan & Kastner, 1996), and $BaSO_4$ exhibits a high fidelity for several other tracers, including the isotopic composition of elements that substitute for Ba (e.g., Ca, Sr), as well as for the mass-dependent and mass-independent isotopic compositions of sulfur and oxygen in the SO_4 portion. The systematics of these substituted and sulfate isotope systems are discussed elsewhere in this series (e.g., Bao et al., 2019; Yao et al., 2020). Below, we discuss putative controls on the fidelity of $\delta^{138}Ba_{barite}$, and highlight other promising archives of $\delta^{138}Ba$.

In the modern ocean, most $BaSO_4$ deposition occurs on the shelf. However, most long-term preservation occurs in the deep ocean (e.g., Paytan & Kastner, 1996; McManus et al., 1998). The proximal reason for this pattern concerns porewater SO_4^{2-} concentrations, which are influenced, among other factors, by organic carbon rain rate. In margin sediments with high organic carbon fluxes, heterotrophic respiration of organic matter near-quantitatively consumes O_2 from porewaters, requiring microbial metabolisms to use alternative electron acceptors such as nitrate, manganese- and iron-oxide minerals, and eventually SO_4^{2-} (e.g., Froelich et al., 1979). Barite preservation requires porewaters to maintain $\Omega_{barite} \geq 1$. A lowering of ambient $[SO_4^{2-}]$ will lower Ω_{barite}, which will be compensated by $BaSO_4$ dissolution. If SO_4^{2-} is near-quantitatively consumed, all $BaSO_4$ will eventually dissolve. Ionic Ba^{2+} will then diffuse within the sediment column until it encounters SO_4^{2-}, at which point $BaSO_4$ will reprecipitate. This process often leads to the development of sharp $BaSO_4$ "fronts" in margin sediments (Torres et al., 1996), which are common at the sulfate–methane transition zone (e.g., Dickens et al., 2003). The consequences of diagenetic $BaSO_4$ cycling on $\delta^{138}Ba_{barite}$ are unknown, but are likely to be significant for two reasons. First, van Zuilen et al. (2016a) reported that diffusive transport of Ba through a silica hydrogel can result in large Ba-isotopic gradients exceeding 1‰. If such findings apply to marine environments, Ba remobilization could be a significant control on sedimentary $\delta^{138}Ba$.

Second, even without transport effects, the combination of dissolution and partial reprecipitation could render large changes in $\delta^{138}Ba_{barite}$ by progressively "distilling" dissolved Ba from porewaters. A potential upside to these complications is that extreme isotopic compositions may be diagnostic of $BaSO_4$ that have been strongly influenced by diagenesis (Section 5).

The cycling of Ba in margin settings means that most non-diagenetically altered $BaSO_4$ is preserved in the deep ocean. In turn, this pattern of preservation implies that little of the pelagic $BaSO_4$ formed prior to ~200 Ma remains, since the deep oceanic crust onto which these precipitates were deposited has been recycled via subduction. Nonetheless, there are significant $BaSO_4$ deposits formed prior to ~200 Ma whose genesis is oftentimes enigmatic (e.g., Hanor, 2000; Horner et al., 2017; Crockford et al., 2019a; Section 4.). As stated above, extreme Ba-isotopic values in such deposits may provide a clear indicator of diagenetic processes, such as multiple $BaSO_4$ dissolution and reprecipitation events. The geological sparsity of these older $BaSO_4$ deposits coupled to the lack of pelagic $BaSO_4$ preservation prior to 200 Ma will necessitate the use of other archives of marine $\delta^{138}Ba_{dBa}$. One promising avenue is marine carbonates. While an experimental study suggests a strong dependence of $\Delta^{138}Ba_{carbonate-dBa}$ on precipitation rate (Mavromatis et al., 2020), studies of Ba isotope systematics in field samples – surface (Liu et al., 2019) and deep-sea corals (Hemsing et al., 2018; Geyman et al., 2019) – suggest that biogenic carbonates can be developed into faithful archives of past $\delta^{138}Ba_{dBa}$. As with $BaSO_4$, however, the fidelity of carbonate $\delta^{138}Ba$ records that have undergone varying degrees of diagenesis is currently unknown (Section 5).

3.2 Measurement

Measurement of $\delta^{138}Ba$ in geologic materials is a three-step process. First, samples must be dissolved. Second, Ba is purified from the sample matrix using ion-exchange chromatography. Third, $\delta^{138}Ba$ is measured using a multi-collector mass spectrometer. Although not essential, it is advantageous to add an isotopic double spike between the first and second steps. We outline these three steps below, and highlight Ba-isotopic results for a $BaSO_4$ international reference material.

If wishing to measure $\delta^{138}Ba_{barite}$ in sediment samples, it is possible to isolate a $BaSO_4$ fraction – typically no more than a few weight percent – from the bulk sediment using either a sequential extraction protocol (e.g., Eagle et al., 2003) or a dispersant (e.g., Tian et al., 2020). It is also possible to measure Ba in bulk sediments and make a correction for lithogenic Ba based on an assumed Ba:Al or Ba:Ti ratio (e.g., Serno et al., 2014; Bridgestock et al., 2018). While bulk approaches will circumvent the process of sequential extraction, they are fraught when the lithogenic correction is large. In contrast, sequential extraction protocols isolate $BaSO_4$ from the sediment matrix. If performing a chemical extraction, Ba that is not associated with $BaSO_4$ is removed from the sample by exposing a suspension of sediment to a series of chemicals tuned to dissolve specific phases (e.g., carbonates, organic matter, opal, and iron-oxide minerals; Eagle et al., 2003). Alternatively, $BaSO_4$ may be isolated by adding a dispersant to a sediment matrix and extracting a suspension containing the $BaSO_4$ fraction (e.g., Tian et al., 2020). Regardless of the extraction method, the resultant $BaSO_4$-rich residue is dissolved separately. The most widely used dissolution methods are multi-acid attack (with or without high pressures), resin exchange, and ligand replacement (e.g., Paytan et al., 1993). Multi-acid attack is not recommended for dissolving $BaSO_4$ isolates since Ba and SO_4 will interact in solution, leading to reprecipitation of $BaSO_4$. Dissolution via resin exchange avoids this issue, although Ba recoveries are rarely quantitative. Ligand replacement can be achieved at high pH using either strong chelating agents (e.g., ethylenediaminetetraacetic- or pentetic acids), or carbonate-for-sulfate substitution via fluxing with Na_2CO_3 (sodium carbonate; Breit et al., 1985). Although slower than the other dissolution methods, the Na_2CO_3 method is among the simplest and does not induce appreciable Ba isotope fractionation at Ba recoveries exceeding 50% (van Zuilen et al., 2016b).

Following dissolution, it is advantageous to add an isotopic double spike to samples. Doing so avoids any fractionation effects associated with Ba loss during subsequent processing steps, and can improve measurement precision, particularly when sample limited. A variety of double spike combinations are described in the literature, with the first reported implementation being [134]Ba–[137]Ba (Eugster et al., 1969). The variety of double spike combinations reflects, in part, the various analytical tradeoffs that must be considered when implementing Ba isotope protocols on a particular mass spectrometer (e.g., overall sensitivity, dynamic range, spectral overlaps). Regardless of whether one uses a double spike, it is necessary to perform ion-exchange chromatography to isolate chemically "pure" Ba for isotopic analysis. Purification is most commonly performed by passing samples through cation-exchange resin in hydrochloric acid. The protocols have remained largely unchanged since the 1970s (e.g., Chan et al., 1977; Horner et al., 2015).

Obtaining high-precision Ba-isotopic data requires analysis via a multi-collector mass spectrometer. The majority of the Ba-isotopic data in the literature were acquired using multi-collector inductively coupled plasma mass spectrometers (MC-ICP-MS), although some groups opted for multi-collector thermal ionization mass spectrometry (MC-TI-MS; e.g., Bullen & Chadwick, 2016;

Hsieh & Henderson, 2017). When employing a double spike and not sample limited, analytical precision for both ionization sources is similar, between ± 0.03–0.05‰ (2σ). More importantly, a number of seawater (Figure 1) and sediment (van Zuilen et al. 2016b) reference materials have been analyzed via both methods and in several labs, yielding highly comparable – and presumably accurate – results. The most widely reported $BaSO_4$ reference material is NBS-127, a powder standard produced by the National Bureau of Standards (now NIST), which possesses $\delta^{138}Ba = -0.28 \pm 0.01$‰ (arithmetic mean ± 2σ of average values reported by Horner et al., 2017; Crockford et al., 2019a; Tian et al., 2020; Tieman et al., 2020). We recommend reporting Ba-isotopic data for NBS-127 and other representative reference materials in publications as a means to ensure data accuracy among the community.

4 Case Studies

Here we highlight three applications of Ba isotopes to study ancient $BaSO_4$ that have provided novel insights into Earth's history. The first two case studies concern the use of Ba isotopes to deduce the origin of enigmatic $BaSO_4$ deposits. The third case study highlights the first example of using Ba isotopes as a tracer of marine biogeochemistry, in this instance across the Paleocene–Eocene Thermal Maximum (PETM).

4.1 Deducing the Origin of Enigmatic Sulfates

4.1.1 Transitioning Out of Earth's Great Oxidation Event

A challenge in interpreting secondary or diagenetic $BaSO_4$ accumulations, particularly in the ancient sedimentary record, is deciphering an approximate age of formation and thus how the geochemistry of these deposits relates to contemporaneous environmental conditions. Recent work from the Paleoproterozoic Belcher group (Costello Fm.) of subarctic Canada provides a useful illustration. Centimeter-sized diagenetic $BaSO_4$ crystals from the Costello Fm. exhibit large negative mass-independent oxygen isotope anomalies (Hodgskiss et al., 2019). A major obstacle to their interpretation, however, concerned the age of the SO_4 that originally bore the large mass-independent oxygen isotope signatures. Possible ages ranged from immediately after Earth's Great Oxidation Event (GOE; ~2.4–2.0 Ga) to several hundred million years later. Hodgskiss et al. (2019) measured $\delta^{138}Ba_{barite}$ in these samples, finding only limited variation about a mean value of ≈ +0.1‰, similar to modern pelagic $BaSO_4$. Based on the narrow range and pelagic-like $\delta^{138}Ba_{barite}$, the authors concluded that the source of Ba to the diagenetic deposits was most likely contemporaneous seawater (Figure 5). Since the fluids transporting Ba and SO_4 that gave rise to these

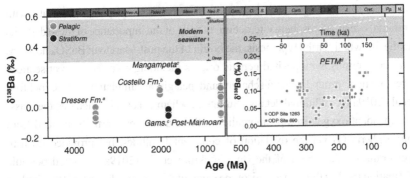

Figure 5 | The record of barium isotopes in barite. Data are plotted against time, with modern and recent samples excluded (see Figure 2). Horizontal shading indicates the modern range of $\delta^{138}Ba_{dBa}$ in seawater. Precambrian samples are separated into their suggested genetic origins of either pelagic or stratiform. Superscripts denote: (a) Horner et al., (2017); (b) Hodgskiss et al., (2019); (c) Crockford et al., (2019a); (d) Bridgestock et al., (2019; see inset).

BaSO$_4$ must have remained segregated until precipitation (Section 2.2.3), these Ba-isotopic data imply that the SO$_4$ must have come from a source other than seawater. Hodgskiss et al. (2019) contend that a likely candidate for the SO$_4$ source was the gypsum pseudomorph-rich Kasegalik Fm., which sits lower in the stratigraphy of the Belcher Group, and was deposited immediately after the GOE around $\approx$ 2,017 Ma. This insight implied that the mass-independent oxygen isotope signatures observed in Costello Fm. BaSO$_4$ were formed immediately after the GOE, evidencing a rapid and major contraction of the biosphere (Hodgskiss et al., 2019) as well as showcasing the utility of $\delta^{138}Ba_{barite}$ to augment findings from other geochemical tracers.

4.1.2 The Marinoan Snowball Earth

Early Ediacaran stratigraphy records the exit from Earth's final Snowball Earth glaciation – the Marinoan ($\approx$ 645[$\pm$ 5]–635 Ma; Hoffman et al., 2017). Within this stratigraphy, certain locations preserve discrete BaSO$_4$ horizons up to tens of cm in thickness (e.g., Bao et al., 2008; Crockford et al., 2018b). The existence of a globally expressed BaSO$_4$ horizon at a specific interval in the sedimentary record is unprecedented, and speaks to a global driving mechanism. These BaSO$_4$ horizons preserve large mass-independent oxygen isotope anomalies that provide some of the most compelling evidence for the Snowball Earth Hypothesis (Hoffman et al., 2017); these anomalies imply extremely elevated atmospheric CO$_2$ levels at the glacial termination (Bao et al., 2008), which is difficult to reconcile with any glacial scenario other than that of a panglacial state with minimal ocean–atmosphere interaction. In spite of the importance of these deposits to the Snowball Earth Hypothesis, the origin of the post-Marinoan $BaSO_4$ horizons was contentious, with posited genetic origins including cold seep, direct seafloor precipitation, groundwater discharge, and pelagic accumulation (see Crockford et al., 2019a for additional details). Sulfate geochemical data implied that the mass-independent oxygen isotope anomalies in these units were partially sourced from atmospheric O_2 via the oxidation of continental sulfides, but provided no firm constraint on the origin of the Ba. Crockford et al., (2019a) explored possible formation models through a Ba-isotope study of four $BaSO_4$ horizons deposited on what is today northwest Canada, south China, Norway, and Brazil. These authors found that all deposits exhibited virtually identical $\delta^{138}Ba_{barite}$ that was similar to modern pelagic $BaSO_4$ (i.e., $\approx$ +0.1‰; Figure 5), implying a common Ba source to all samples. The authors concluded that the most likely common Ba source was contemporaneous seawater, as both the value and range of Ba isotope compositions was incompatible with either a cold seep or groundwater origin, respectively. Crockford et al. (2019a) contended that post-Marinoan $BaSO_4$ formed as a direct seafloor precipitate at the interface between a Ba-rich deep ocean and highly stratified, sulfate-rich, post-glacial surface reservoir, and that the glaciation allowed marine [Ba] to reach concentrations not seen in the ~650 Myr since. Together with the results highlighted above from the Belcher Group, these examples showcase how $\delta^{138}Ba_{barite}$ can be used to explore the genetic origins of $BaSO_4$ deposits older than 200 Myr.

4.2 Tracing Marine Biogeochemistry

The potential for Ba-isotopes to provide novel insights into marine biogeochemistry is the second major thrust of this emerging tracer. The first example of this application is found in Bridgestock et al. (2019). These authors examined a major, albeit brief (< 50 kyr), climate warming event $\approx$ 55.5 Ma – the PETM. The initial warming is thought to have been driven by rapid carbon injection into the ocean–atmosphere system. However, the mechanisms driving the increased carbon sequestration during the climate recovery remain contentious, with explanations ranging from enhanced silicate weathering to increased organic carbon export. Observations of increased sedimentary $BaSO_4$ accumulation are consistent with the latter explanation, although these records are inherently local and do not rule out decreases in export in other parts of the ocean or transient enhancement of Ba burial driven by increased Ba inputs (e.g., Dickens et al., 2003). Recognizing this limitation, Bridgestock et al. (2019) analyzed the $\delta^{138}Ba$ of bulk sediment from two cores, one in the South Atlantic and another from the Southern Ocean. The authors show that the cycling of Ba is relatively unchanged during the onset of the PETM, which is interpreted as evidencing only minor changes in export production. During the PETM recovery, when $BaSO_4$ accumulation rates increase, $\delta^{138}Ba$ also increases (Figure 5). The authors contend that these correlated increases are the result of increased export production, rather than changes in Ba inputs or other artefacts, such as from dilution or poor age models. Bridgestock et al. (2019) thus conclude that the PETM climate recovery was aided by increased organic carbon burial. This example illustrates how sedimentary Ba isotope measurements can be used to refine interpretations made using other productivity tracers. Likewise, these data illustrate how $\delta^{138}Ba_{barite}$ depends on multiple processes, and deriving unique geological interpretations requires multiple contemporaneous records and careful accounting of other processes that might impact sedimentary $\delta^{138}Ba$ (Section 5.).

5 Future Prospects

This study summarizes the first decade of stable Ba isotope geochemistry, which has revealed a number of global, regional, local, and geological processes that can render different fluid reservoirs with distinct $\delta^{138}Ba_{dBa}$. Since the isotopic composition of resultant $BaSO_4$, $\delta^{138}Ba_{barite}$, depends largely on the quantity and isotopic composition of Ba in the starting reservoir, this tracer can be used to glean unique insights into the sources and cycling of Ba in the environment. Despite recent progress, there remain a number of key uncertainties in the cycling of Ba that require resolution, including: the formation of $BaSO_4$ itself, the marine budget of Ba and its isotopes, and challenges in constraining the cycling of Ba back through time.

Regarding $BaSO_4$ formation, it is imperative that the discrepancy between lab- and field-based estimates of $\Delta^{138}Ba_{barite-dBa}$ be resolved. It is possible that this discrepancy reflects the lack of laboratory $BaSO_4$ precipitation experiments conducted under marine-analogue conditions, specifically those that systematically explore the role of precipitation rate, salinity, trace element content, and pressure. Alternatively, some of the discrepancy may derive from processes involving organic matter. Addressing this point requires studying particulate Ba isotopes in the marine realm on scales ranging from microenvironments to ecosystems and answering three questions: how much Ba in $BaSO_4$ is sourced from organic matter versus ambient seawater; are there dependencies on ecology (plankton assemblage, particle size and aggregation, biofilm production); and, do other environmental factors play a role (e.g., $[O_2]$, Ω_{barite})? The answers to all three questions are largely unknown and their answers have likely changed over time. More modern process studies in diverse environments are thus essential for refining the utility of $\delta^{138}Ba$.

The marine Ba budget is insufficiently constrained. The modern Ba cycle is either not at steady state, or is missing at least one major flux term possessing $\delta^{138}Ba \leq +0.1‰$. Continued efforts are needed to accurately constrain the identity, magnitude, and Ba-isotopic composition of this flux. Surveys of $\delta^{138}Ba$ in estuaries, hydrothermal systems, cold seeps, and continental margins will help narrow the search. The magnitude and Ba-isotopic composition of riverine Ba also remains relatively uncertain. Given that silicate weathering on land is the principal Ba source to seawater, it will be necessary to consider how riverine Ba fluxes might vary under different continental configurations, climate states, and terrestrial biospheres. Likewise, changing seawater chemistry may have modified the importance of certain intra-marine Ba sources; lower marine $[SO_4^{2-}]$ could enhance large Ba point sources and potentially increase the importance of other dissolved–particulate Ba transformations that are less significant today (e.g., interactions with iron oxides, clay minerals).

Constraining the Ba-isotopic cycle in the past will be the most challenging task in terms of assessing fidelity and interpreting accurately. The fidelity of $BaSO_4$-hosted $\delta^{138}Ba$ records is almost entirely unconstrained during early diagenesis. On long timescales, stable mineral recrystallization may be significant. Investigating potential sensitivities will require studies of co-located $BaSO_4$ and porewaters. Additionally, it may be beneficial to investigate other archives of the marine Ba cycle, such as carbonates. Results for modern and recent samples are promising, although, again, the role of early diagenesis is largely unknown. Lessons learned from other Group II isotope systems offer a road map for developing $\delta^{138}Ba$ in marine carbonates. Lastly, it is imperative that the drivers of vertical and spatial heterogeneity in $\delta^{138}Ba_{dBa}$ be considered when interpreting sedimentary records. On a global basis, the vertical gradient in $\delta^{138}Ba_{dBa}$ depends on both biological productivity and ocean circulation. Regional circulation patterns and "upstream" processes have a similarly profound effect on local $\delta^{138}Ba_{dBa}$. As such, no individual record of $\delta^{138}Ba_{dBa}$ recovered from a single site can be unambiguously interpreted in terms of either biological or physical processes without additional constraints on either productivity or circulation. Far from being a weakness, this aspect of Ba isotope geochemistry may prove decisive in differentiating local versus regional changes in carbon export and ocean circulation through time. Indeed, the next decade of Ba isotope geochemistry is poised to offer many novel insights into the origins of enigmatic $BaSO_4$ deposits, the transport and cycling of Ba in the environment, and Earth's biogeochemical evolution.

6 Key Papers

Demonstration of Mass-Dependent Ba-Isotopic Variations in the Environment

von Allmen et al. (2010) were the first to report experimental constraints on the Ba-isotopic fractionation factor for $BaSO_4$ precipitation. Horner et al. (2015) were the first to precisely and accurately report a Ba-isotopic profile in seawater.

von Allmen, K., Böttcher, M. E., Samankassou, E., and Nägler, T. F. (2010). Barium isotope fractionation in the global barium cycle: First evidence from barium minerals and precipitation experiments. *Chemical Geology*, **277**(1–2), 70–7.

Horner, T. J., Kinsley, C. W., and Nielsen, S. G. (2015). Barium-isotopic fractionation in seawater mediated by barite cycling and oceanic circulation. *Earth and Planetary Science Letters*, **430**, 511–22.

Barium-Isotopic Studies in Marine Sediments

These studies illustrate the two major applications of Ba isotopes, as reviewed in Section 4. Bridgestock et al. (2019) apply Ba isotopes as a proxy for marine biogeochemistry across a major climatic event. Crockford et al. (2019a) used Ba isotopes to constrain the origin of enigmatic $BaSO_4$ deposited after the Marinoan Snowball Earth, and included Ba isotope data for scores of modern core-top samples.

Bridgestock, L., Hsieh, Y. T., Porcelli, D., and Henderson, G. M. (2019). Increased export production during recovery from the Paleocene–Eocene thermal maximum constrained by sedimentary Ba isotopes. *Earth and Planetary Science Letters*, **510**, 53–63.

Crockford, P. W., Wing, B. A., Paytan, A. et al. (2019a). Barium-isotopic constraints on the origin of post-Marinoan barites. *Earth and Planetary Science Letters*, **519**, 234–44.

The Particulate Ba Cycle in the Open Ocean

Dehairs et al. (1980) demonstrated that the vast majority of particulate Ba in the water column was microcrystalline $BaSO_4$ and associated with biological debris. Bishop (1988) identified that pelagic $BaSO_4$ formed within large aggregates and that the μm-scale precipitates were released to seawater during disaggregation.

Dehairs, F., Chesselet, R., and Jedwab, J. (1980). Discrete suspended particles of barite and the barium cycle in the open ocean. *Earth and Planetary Science Letters*, **49**(2), 528–50.

Bishop, J. K. (1988). The barite-opal-organic carbon association in oceanic particulate matter. *Nature*, **332**(6162), 341.

Geological Controls on Ba Abundances in Seawater and Sediments

Walker (1983) used saturation state modeling to suggest that the barium and sulfate ion products have not appreciably changed over time, implying that dissolved Ba was much higher in the Archaean ocean. Hanor (2000) provides a veritable tour de force that reviews the geochemistry, occurrence, and significance of $BaSO_4$ in the geological record. Dickens et al. (2003) developed an elegant numerical framework for describing the relationship between Ba fluxes and productivity on a global basis.

Walker, J. C. (1983). Possible limits on the composition of the Archaean ocean. *Nature*, **302**(5908), 518–20.

Hanor, J. S. (2000). Barite–celestine geochemistry and environments of formation. *Reviews in Mineralogy and Geochemistry*, **40**(1), 193–275.

Dickens, G. R., Fewless, T., Thomas, E., and Bralower, T. J. (2003). Excess barite accumulation during the Paleocene-Eocene Thermal Maximum: Massive input of dissolved barium from seafloor gas hydrate reservoirs. In Wing, S. L, Gingerich, P. D., Schmitz, B., and Thomas, E. (Eds.) *Causes and consequences of globally warm climates in the early Paleogene*, **369**, 11–23, Geological Society of America Special Papers.

References

Bao, H. (2019). Triple Oxygen Isotopes (Elements in Geochemical Tracers in Earth System Science). Cambridge: Cambridge University Press.

Bao, H., Lyons, J. R., and Zhou, C. (2008). Triple oxygen isotope evidence for elevated CO_2 levels after a Neoproterozoic glaciation. *Nature*, **453**(7194), 504–6.

Bates, S. L., Hendry, K. R., Pryer, H. V. et al. (2017). Barium isotopes reveal role of ocean circulation on barium cycling in the Atlantic. *Geochimica et Cosmochimica Acta*, **204**, 286–299.

Blättler, C.L. and Higgins, J.A., (2014). Calcium isotopes in evaporites record variations in Phanerozoic seawater SO4 and Ca. *Geology*, **42**(8), 711–14.

Böttcher, M. E., Neubert, N., Von Allmen, K., Samankassou, E. and Nägler, T. F. (2018). Barium isotope fractionation during the experimental transformation of aragonite to witherite and of gypsum to barite, and the effect of ion (de) solvation. *Isotopes in Environmental and Health Studies*, **54**(3), 324–35.

Breit, G. N., Simmons, E. C., and Goldhaber, M. B. (1985). Dissolution of barite for the analysis of strontium isotopes and other chemical and isotopic variations using aqueous sodium carbonate. *Chemical Geology: Isotope Geoscience Section*, **52**(3–4), 333–6.

Bridgestock, L., Hsieh, Y. T., Porcelli, D. et al. (2018). Controls on the barium isotope compositions of marine sediments. *Earth and Planetary Science Letters*, **481**, 101–10.

Bullen, T. and Chadwick, O. (2016). Ca, Sr and Ba stable isotopes reveal the fate of soil nutrients along a tropical climosequence in Hawaii. *Chemical Geology*, **422**, 25–45.

Cao, Z., Li, Y., Rao, X. et al. (2020a). Constraining barium isotope fractionation in the upper water column of the South China Sea. *Geochimica et Cosmochimica Acta*, **288**, 120–37.

Cao, Z., Siebert, C., Hathorne, E. C., Dai, M., and Frank, M. (2020b). Corrigendum to "Constraining the oceanic barium cycle with stable barium isotopes" [Earth Planet. Sci. Lett. 434 (2016) 1–9], *Earth and Planetary Science Letters*, **530**, 116003.

Chan, L. H., Drummond, D., Edmond, J. M., and Grant, B. (1977). On the barium data from the Atlantic GEOSECS expedition. *Deep Sea Research*, **24**(7), 613–49.

Charbonnier, Q., Bouchez, J., Gaillardet, J., and Gayer, É. (2020). Barium stable isotopes as a fingerprint of biological cycling in the Amazon River Basin. *Biogeosciences*, **17**(23), 5989–6015.

Chow, T. J. and Goldberg, E. D. (1960). On the marine geochemistry of barium. *Geochimica et Cosmochimica Acta*, **20**(3–4), 192–8.

Church, T. M. and Wolgemuth, K. (1972). Marine barite saturation. *Earth and Planetary Science Letters*, **15**(1), 35–44.

Crockford, P. W., Hayles, J. A., Bao, H. et al. (2018a). Triple oxygen isotope evidence for limited mid-Proterozoic primary productivity. *Nature*, **559** (7715), 613–16.

Crockford, P. W., Hodgskiss, M. S. W., Uhlein, G. J. et al. (2018b). Linking paleocontinents through triple oxygen isotope anomalies. *Geology*, **46**(2), 179–82.

Crockford, P. W., Kunzmann, M., Bekker, A. et al. (2019b). Claypool continued: Extending the isotopic record of sedimentary sulfate. *Chemical Geology*, **513**, 200–25.

Dempster, A. J. (1936). The Isotopic Constitution of Barium and Cerium. *Physical Review*, **49**(12), 947.

Deng, N., Stack, A. G., Weber, J., Cao, B., De Yoreo, J. J., and Hu, Y. (2019). Organic–mineral interfacial chemistry drives heterogeneous nucleation of Sr-rich (Ba_x, Sr_{1-x})SO_4 from undersaturated solution. *Proceedings of the National Academy of Sciences*, **116**(27), 13221–6.

Eagle, M., Paytan, A., Arrigo, K. R., van Dijken, G., and Murray, R. W. (2003). A comparison between excess barium and barite as indicators of carbon export. *Paleoceanography*, **18**(1).

Eugster, O., Tera, F., and Wasserburg, G. J. (1969). Isotopic analyses of barium in meteorites and in terrestrial samples. *Journal of Geophysical Research*, **74**(15), 3897–908.

Falkner, K. K., Bowers, T. S., Todd, J. F. et al. (1993). The behavior of barium in anoxic marine waters. *Geochimica et Cosmochimica Acta*, **57**(3), 537–54.

Froelich, P., Klinkhammer, G. P., Bender, M. L. et al. (1979). Early oxidation of organic matter in pelagic sediments of the eastern equatorial Atlantic: suboxic diagenesis. *Geochimica et Cosmochimica Acta*, **43**(7), 1075–90.

Ganeshram, R. S., François, R., Commeau, J., and Brown-Leger, S. L. (2003). An experimental investigation of barite formation in seawater. *Geochimica et Cosmochimica Acta*, **67**(14), 2599–605.

Geyman, B. M., Ptacek, J. L., LaVigne, M., and Horner, T. J., (2019). Barium in deep-sea bamboo corals: Phase associations, barium stable isotopes, & prospects for paleoceanography. *Earth and Planetary Science Letters*, **525**, 115751.

Gong, Y., Zeng, Z., Cheng, W. et al. (2020). Barium isotopic fractionation during strong weathering of basalt in a tropical climate. *Environment International*, **143**, 105896.

Gou, L. F., Jin, Z., Galy, A. et al. (2020). Seasonal riverine barium isotopic variation in the middle Yellow River: Sources and fractionation. *Earth and Planetary Science Letters*, **531**, 115990.

Gou, H., Li, W. Y., Nan, X., and Huang, F. Experimental *evidence for light Ba isotopes favouring aqueous fluids over silicate melts. Geochemical Perspectives Letters*, 16, 6–11.

Hemsing, F., Hsieh, Y. T., Bridgestock, L. et al. (2018). Barium isotopes in cold-water corals. *Earth and Planetary Science Letters*, **491**, 183–92.

Hodgskiss, M. S., Crockford, P. W., Peng, Y., Wing, B. A., and Horner, T. J. (2019). A productivity collapse to end Earth's Great Oxidation. *Proceedings of the National Academy of Sciences*, **116**(35), 17207–12.

Hoffman, P. F., Abbot, D. S., Ashkenazy, Y. et al. (2017). Snowball Earth climate dynamics and Cryogenian geology-geobiology. *Science Advances*, *3*(11), e1600983.

Hoppema, M., Dehairs, F., Navez, J. et al. (2010). Distribution of barium in the Weddell Gyre: Impact of circulation and biogeochemical processes. *Marine Chemistry*, **122**(1–4), 118–29.

Horner, T. J., Pryer, H. V., Nielsen, S. G. et al. (2017). Pelagic barite precipitation at micromolar ambient sulfate. *Nature Communications*, **8**(1), 1342.

Horner, T. J., Rickaby, R.E., and Henderson, G.M. (2011). Isotopic fractionation of cadmium into calcite. *Earth and Planetary Science Letters*, **312**(1–2), 243–53.

Hsieh, Y. T., Bridgestock, L., Scheuermann, P. P., Seyfried Jr, W. E., and Henderson, G. M. (2021). Barium isotopes in mid-ocean ridge hydrothermal vent fluids: A source of isotopically heavy Ba to the ocean. *Geochimica et Cosmochimica Acta*, **292**, 348–63.

Hsieh, Y. T., and Henderson, G. M. (2017). Barium stable isotopes in the global ocean: Tracer of Ba inputs and utilization. *Earth and Planetary Science Letters*, **473**, 269–78.

Jacquet, S. H., Dehairs, F., Elskens, M., Savoye, N., and Cardinal, D., (2007). Barium cycling along WOCE SR3 line in the Southern Ocean. *Marine Chemistry*, **106**(1–2), 33–45.

Li, W. Y., Yu, H. M., Xu, J. et al. (2020). Barium isotopic composition of the mantle: Constraints from carbonatites. *Geochimica et Cosmochimica Acta*, **278**, 235–43.

Li, Y. H., and Chan, L. H. (1979). Desorption of Ba and ^{226}Ra from river-borne sediments in the Hudson estuary. *Earth and Planetary Science Letters*, **43**(3), 343–50.

Liu, Y., Li, X., Zeng, Z. et al. (2019). Annually-resolved coral skeletal $\delta^{138/134}$Ba records: A new proxy for oceanic Ba cycling. *Geochimica et Cosmochimica Acta*, **247**, 27–39.

Martínez-Ruiz, F., Paytan, A., Gonzalez-Muñoz, M. T. et al. (2019). Barite formation in the ocean: Origin of amorphous and crystalline precipitates. *Chemical Geology*, **511**, 441–51.

Martínez-Ruiz, F., Paytan, A., Gonzalez-Muñoz, M. T. et al. (2020). Barite precipitation on suspended organic matter in the mesopelagic zone. *Frontiers in Earth Science: Biogeosciences*, **8**, 499–515.

Mavromatis, V., van Zuilen, K., Blanchard, M et al. (2020). Experimental and theoretical modelling of kinetic and equilibrium Ba isotope fractionation during calcite and aragonite precipitation. *Geochimica et Cosmochimica Acta*, **269**, 566–80.

Mayfield, K. K., Eisenhauer, A., Santiago Ramos, D., et al. (2021) Groundwater discharge impacts marine isotope budgets of Li, Mg, Ca, Sr, and Ba, *Nature Communications*, **12**, 148.

McManus, J., Berelson, W. M., Klinkhammer, G. P., Kilgore, T. E., and Hammond, D. E. (1994). Remobilization of barium in continental margin sediments. *Geochimica et Cosmochimica Acta*, **58**(22), 4899–907.

McManus, J., Berelson, W. M., Klinkhammer, G. P. et al. (1998). Geochemistry of barium in marine sediments: Implications for its use as a paleoproxy. *Geochimica et Cosmochimica Acta*, **62**(21–2), 3453–73.

Monnin, C., Jeandel, C., Cattaldo, T., and Dehairs, F. (1999). The marine barite saturation state of the world's oceans. *Marine Chemistry*, **65**(3–4), 253–61.

Nan, X. Y., Yu, H. M., Rudnick, R. L. et al. (2018). Barium isotopic composition of the upper continental crust. *Geochimica et Cosmochimica Acta*, **233**, 33–49.

Nancollas, G. H. and Purdie, N. (1963). Crystallization of barium sulphate in aqueous solution. *Transactions of the Faraday Society*, **59**, 735–40.

Nielsen, S. G., Horner, T. J., Pryer, H. V. et al. (2018). Barium isotope evidence for pervasive sediment recycling in the upper mantle. *Science Advances*, **4** (7), eaas8675.

Nielsen, S. G., Shu, Y., Auro, M. et al. (2020). Barium isotope systematics of subduction zones. *Geochimica et Cosmochimica Acta*, **275**, 1–18.

Parkhurst, D. L. and Appelo, C. A. J. (1999). User's guide to PHREEQC (Version 2): A computer program for speciation, batch-reaction, one-dimensional transport, and inverse geochemical calculations. *Water-Resources Investigations Report*, **99**(4259), 312.

Paytan, A. and Griffith, E. M. (2007). Marine barite: Recorder of variations in ocean export productivity. *Deep Sea Research Part II: Topical Studies in Oceanography*, **54**(5–7), 687–705.

Paytan, A. and Kastner, M. (1996). Benthic Ba fluxes in the central Equatorial Pacific, implications for the oceanic Ba cycle. *Earth and Planetary Science Letters*, **142**(3–4), 439–50.

Paytan, A., Kastner, M., Martin, E. E., Macdougall, J. D., and Herbert, T. (1993). Marine barite as a monitor of seawater strontium isotope composition. *Nature*, **366**(6454), 445–9.

Paytan, A., Mearon, S., Cobb, K., and Kastner, M. (2002). Origin of marine barite deposits: Sr and S isotope characterization. *Geology*, **30**(8), 747–50.

Rushdi, A. I., McManus, J., and Collier, R. W. (2000). Marine barite and celestite saturation in seawater. *Marine Chemistry*, **69**(1–2), 19–31.

Serno, S., Winckler, G., Anderson, R. F. et al. (2014). Using the natural spatial pattern of marine productivity in the Subarctic North Pacific to evaluate paleoproductivity proxies. *Paleoceanography*, **29**(5), 438–53.

Talley, L. D. (2013). Closure of the global overturning circulation through the Indian, Pacific, and Southern Oceans: Schematics and transports. *Oceanography*, **26**(1), 80–97.

Tian, L. L., Gong, Y. Z., Wei, W. et al. (2020). Rapid determination of Ba isotope compositions for barites using a H_2O-extraction method and MC-ICP-MS. *Journal of Analytical Atomic Spectrometry*, **35**(8), 1566–73.

Tieman, Z. G., Stewart, B. W., Capo, R. C. et al. (2020). Barium isotopes track the source of dissolved solids in produced water from the unconventional Marcellus Shale Gas Play. *Environmental Science & Technology*, **54**(7), 4275–85.

Torres, M. E., Bohrmann, G., Dubé, T. E., and Poole, F. G. (2003). Formation of modern and Paleozoic stratiform barite at cold methane seeps on continental margins. *Geology*, **31**(10), 897–900.

Torres, M. E., Brumsack, H. J., Bohrmann, G., and Emeis, K. C. (1996). Barite fronts in continental margin sediments: a new look at barium remobilization in the zone of sulfate reduction and formation of heavy barites in diagenetic fronts. *Chemical Geology*, **127**(1–3), 125–39.

van Beek, P., Francois, R., Conte, M. et al. (2007). $^{228}Ra/^{226}Ra$ and $^{226}Ra/Ba$ ratios in seawater and particles at the OFP site in the western Sargasso Sea near Bermuda. *Geochimica et cosmochimica Acta*, **71**, 71–86.

van Zuilen, K., Müller, T., Nägler, T. F., Dietzel, M., and Küsters, T. (2016a). Experimental determination of barium isotope fractionation during diffusion and adsorption processes at low temperatures. *Geochimica et Cosmochimica Acta*, **186**, 226–41.

van Zuilen, K., Nägler, T. F., and Bullen, T.D. (2016b). Barium isotopic compositions of geological reference materials. *Geostandards and Geoanalytical Research*, **40**(4), 543–58.

Wedepohl, K. H. (1995). The composition of the continental crust. *Geochimica et Cosmochimica Acta*, **59**(7), 1217–32.

Yao, W., Griffith, E., and Paytan, A. (2020). Pelagic Barite: Tracer of Ocean Productivity and a Recorder of Isotopic Compositions of Seawater S, O, Sr, Ca and Ba (Elements in Geochemical Tracers in Earth System Science). Cambridge: Cambridge University Press.

Acknowledgments

The authors thank the members of the NIRVANA Labs, David Johnston, and Adina Paytan for many engaging discussions; Luke Bridgestock for advice on Section 4.2; Wei Wei and an anonymous referee for detailed and constructive feedback; and, the series editors for their patience during these unprecedented times. T. J. H. acknowledges financial support from the US National Science Foundation and P. W. C. from the Agouron Geobiology Postdoctoral Fellowship Program.

Cambridge Elements $\equiv$

Elements in Geochemical Tracers in Earth System Science

Timothy Lyons

University of California

Timothy Lyons is a distinguished professor of biogeochemistry in the Department of Earth Sciences at the University of California, Riverside. He is an expert in the use of geochemical tracers for applications in astrobiology, geobiology and Earth history. Professor Lyons leads the 'Alternative Earths' team of the NASA Astrobiology Institute and the Alternative Earths Astrobiology Center at UC Riverside.

Alexandra Turchyn

University of Cambridge

Alexandra Turchyn is a university reader in biogeochemistry in the Department of Earth Sciences at the University of Cambridge. Her primary research interests are in isotope geochemistry and the application of geochemistry to interrogate modern and past environments.

Chris Reinhard

Georgia Institute of Technology

Chris Reinhard is an assistant professor in the Department of Earth and Atmospheric Sciences at the Georgia Institute of Technology. His research focuses on biogeochemistry and paleoclimatology, and he is an Institutional PI on the 'Alternative Earths' team of the NASA Astrobiology Institute.

About the Series

This innovative series provides authoritative, concise overviews of the many novel isotope and elemental systems that can be used as "proxies" or "geochemical tracers" to reconstruct past environments over thousands to millions to billions of years – from the evolving chemistry of the atmosphere and oceans to their cause-and-effect relationships with life.

Covering a wide variety of geochemical tracers, the series reviews each method in terms of the geochemical underpinnings, the promises and pitfalls, and the "state-of-the-art" and future prospects, providing a dynamic reference resource for graduate students, researchers and scientists in geochemistry, astrobiology, paleontology, paleoceanography and paleoclimatology.

The short, timely, broadly accessible papers provide much-needed primers for a wide audience – highlighting the cutting-edge of both new and established proxies as applied to diverse questions about Earth system evolution over wide-ranging time scales.

Cambridge Elements ☰

Elements in Geochemical Tracers in Earth System Science

Elements in the Series

A full series listing is available at: www.cambridge.org/EESS